Country Garden & Smallholding
Magazine

Broad Leys Publishing Company

Introduction

The Smallholding Plan from *Country Garden & Smallholding* magazine provides a blueprint for anyone about to buy a smallholding or who wishes to make productive use of a little land. This may be anything from a large garden and small orchard to several acres, but whatever the scale, the following factors need to be taken into account: *Space; Time; Energy; Money* (Interest is assumed as a prerequisite!)

Any activity needs space. The size of your garden or amount of land will determine what is practically possible. The Plan looks at how much space different activities need. If you are purchasing a smallholding, and you have an idea what activities you intend to pursue, then this information will be useful.

Everything we do takes time. By considering how 'time-consuming' different activities are the Plan shows how much it is realistic to undertake, given your own limitations in this respect.

Energy is finite. What may be a good idea for the young and fit may prove onerous for those taking early retirement. The Plan considers this aspect, too. It also considers whether the setting up and running costs of various enterprises are high or low. Remember that the different activities do add up, although some may fit well together!

This booklet does not claim to be comprehensive, or to provide a wealth of practical detail. Instead, it concentrates on the salient points of each smallholding activity, whilst pointing the reader in the right direction for further information.

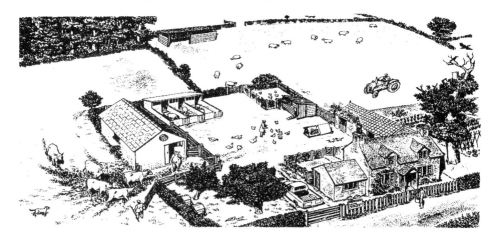

© Country Garden & Smallholding. 1996   ISBN 0 906137 24 1

Published by Country Garden & Smallholding magazine, Broad Leys Publishing Company, Buriton House, Station Road, Newport, Saffron Walden, Essex CB11 3PL.
Tel: 01799 540922 Fax: 01799 541367.

No material may be reproduced without the permission of the publishers,

Contents

Introduction	2
Buying a Country Property	4
Smallholding Checklist	5
The Kitchen Garden	6
Chickens	8
Ducks	10
Geese	11
Turkeys	12
Quail	14
Rabbits	15
Sheep	16
Goats	18
House Cow	20
Pigs	22
Bees	24

All books recommended in *The Smallholding Plan* are available from the *Country Garden & Smallholding Book Catalogue* - see inside back cover.

Buying a Country Property

If buying a country property, do go and look at it in the winter, when there's plenty of mud about. It's easy to be beguiled when the sun is shining and all the cottage flowers are in bloom. There are many other aspects to be considered. Does the property really meet your needs, or can it be adapted at a reasonable cost? Use the checklist on the page opposite, and be honest with your answers!

Once the property is bought, for better or worse, a temptation is to start with livestock too early. It is much better to concentrate on the house and get that repaired, renovated or redecorated, as necessary, before thinking of outside activities. If money is no object, this may not apply, as outside contractors will probably be doing the work, but most of us will be doing our own DIY, and that takes a lot of time and energy.

Gaining experience of smallholding practice and looking after animals *before* starting a practical enterprise is vital. There are many short courses being offered by agricultural colleges, the *Agricultural Training Board*, and many others. Find out if there is a local smallholding group in your area (ask at your local library or *Country Garden & Smallholding* magazine). Such groups often have their own training programmes and are good sources of local information and help.

After the house, the garden is probably next in the order of priorities, with fencing or hedging the boundaries following on. In fact, no livestock at all should be considered before the perimeter fencing is checked and made good. Wandering stock is potentially dangerous on the road, and you could be liable in the event of an accident. Where hedges are overgrown and straggly, trim them back, or layer them, so that they can thicken at the base. Winter is a good time to plant new hedges or replace gaps with new saplings. They will need to be protected with wire netting against stock until they are fully grown and dense.

Outbuildings can usually be restored as and when finance is available, as long as they are not needed sooner for animal housing or feed storage. There is also a whole range of small scale housing available for animals and poultry, as well as plans for the DIY enthusiast.

Now, let's look at each smallholding activity in three ways - the *key factors - space, time, energy and money* requirements - and where to go for *further information.*

Left: Method of layering a hedge to thicken the growth

Question	Answer	Question	Answer
Is the house suitable for you and all your family?		Are there any restrictions on the property?	
Is there mains or private water? What condition?		Mains/private sewage? What condition?	
Mains electricity or private? What condition?		Does the house need to be altered or renovated?	
Problem neighbours - traffic, noise, pollution or other aggravation?		Is there easy vehicular access to the main road?	
Have you seen the property in cold, wet conditions?		Are there any plans for new highways nearby?	
How far are the nearest town, village and shops?		How far is the nearest school?	
How far is the nearest doctor/hospital?		Is there effective public transport available?	
Any outbuildings? What state?		What is the condition of the fences and the gates?	
What is the condition of the hedges and ditches?		Is there a lot of rubbish on the site?	
How high is the property?		Is it on steep ground?	
Is it north or south facing?		Is it sheltered or exposed?	
What are the prevailing weather conditions?		Are the fields of convenient shape and size?	
Is there a garden? What is its condition? What are the prevalent weeds?		What is land drainage like? Rushes and reeds?	
Has a high level of chemicals been applied to the land in the past?		Are there any footpaths or other rights of way?	
What grade agricultural land is it?		What is the pH of the soil?	
What trees are present on the site?		Is the property in a tourist area?	
Is there any evidence of vermin?		Are there any shooting or fishing rights?	
Suitable venues, eg church, youth club, pub?		Looking at your answers, is the property really suitable? (Be honest!)	

The Kitchen Garden

It is assumed that anyone thinking of starting a kitchen garden will be growing fruit and vegetables organically. If it is to be a chemical feast, the produce might as well come from the local supermarket. The whole point of growing your own is that produce is fresher and the quality is higher, without chemical residues.

Key factors

• Prepare the plot thoroughly by digging over and removing perennial weed roots such as docks, nettles and thistles. It may be necessary to scythe or flame-gun first in order to clear the area. Alternatively cover the plot with weighted-down black polythene and leave for 6 months to kill off plant growth underneath. (It is also possible to allow pigs or poultry to prepare the ground).
• Build up soil structure and fertility by adding compost and manure (based on pig, poultry, goat, sheep or rabbit droppings). It may be necessary to buy organic fertilizer until you are in a position to produce your own compost. Spent mushroom compost and horse manure are available in most areas.
• Start a series of compost heaps as soon as possible, but avoid adding perennial weeds.
• Grow only those crops that are suited to the soil. If necessary, test the pH value of the soil with a soil test kit. Lime if necessary.
• Concentrate on crops that the family will eat, or for which there is a local market. These may be luxury items such as asparagus or bunches of early carrots. No produce may be sold with the description 'organic' unless the producer is registered with the *United Kingdom Register of Organic Food Standards (UKROFS)* which provides a set of national standards, and a certification and inspection scheme. (There are also other organisations such as the *Soil Association, Organic Farmers & Growers*, and *Biodynamic Association* which have their own standards, and are registered with *UKROFS*).
• Choose fruit and vegetable varieties that have been developed with some degree of disease/pest

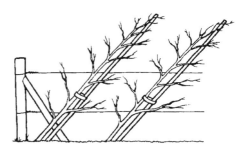

Cordon fruit trees can be grown as a hedge in order to maximise production area. They are tied to canes which are fixed at an angle to horizontal wires.

resistance and which have a good flavour (Check the seed and plant catalogues).
• Raise crops that produce a good yield in relation to the amount of ground used and which cannot be bought cheaply elsewhere, eg, at a local PYO farm. Fruit trees, for example, can be bought on dwarf root stocks or even used as fences if grown on a cordon system.
• Hoe regularly between the crops to keep down annual weeds.
• Mulch around crops to suppress weeds and retain moisture.
• Use a seep hose or trickle system of irrigation which takes water straight to the roots, and avoids surface waste.
• Where there is a problem, use biological controls, rather than chemical methods. There are specialist suppliers.
• Pick regularly to maximise production and eat, freeze or sell as soon as possible.
• Have a rotation system so that crops are not grown on the same area two years running.
• Sow green manures as part of the rotation plan to add to the soil fertility

Space, Time, Energy and Money

Space - Anything from a garden plot to an allotment or field.

Time - Preparation time is high and crops also need constant attention, particularly in summer. Selling produce may require research to find outlets unless it is 'at the gate'.

Energy - Lots needed, particularly for digging heavy soils (beware backache). Tackle what you can, using a step-by-step approach, ie, prepare one bed at a time.

Money - Starting costs can be high if you need to buy in composts, mulches and plants. An irrigation system can be expensive, but is a 'once only' capital cost. Once you are in a position to produce your own compost and raise plants from seed, the running costs are relatively low. If you sell organic produce, the costs of registration and compliance with UKROFS regulations are quite high, and may not be worth it on a small scale.

Further information

The Henry Doubleday Research Association, Ryton Organic Gardens, Ryton-on-Dunsmore, Coventry CV8 3LG. Tel: 01203 303517 . An association for organic gardeners. Gardens that are open to the public.

UKROFS, Room 320C, Nobel House, 17 Smith Square, London SW1P 3JR. Tel: 0171 238 5915. Information on selling organic produce.

Further Reading - The Self-Sufficient Gardener by John Seymour.

Chickens

Chickens are easy to keep and can be kept on a garden, orchard or field scale. In a small garden, bantams may be more appropriate. The easiest way to start is to buy point-of-lay pullets at around 18 weeks. They should start to lay from around 21 weeks. The choice is between *modern hybrids* such as ISA Brown, Hisex Ranger and Lohmann Brown, *commercial first crosses* such as Black Rock, which have all been bred for high egg production, and *traditional pure breeds*, such as Rhode Island Red, Maran or Light Sussex, etc, which will lay less but are popular with those interested in helping to conserve the old breeds.

Most commercial point-of-lay pullets will have been inoculated against a number of diseases such as Newcastle and Mareks diseases. Check with the supplier. Generally, 3-4 good layers provide enough eggs for a family.

Key factors
• Check that there are no local bylaws or restrictive covenants on the property that ban the keeping of poultry.
• Chickens need a dry, well ventilated house, equipped with a perch, pop-hole exit and nest boxes. The latter should be accessible from the outside for ease of egg collection. There should also be a door or other access (such as the roof lifting off) for ease of cleaning. A droppings board or droppings pit under the perch facilitates cleaning.
• They need a run, either attached to the house, or an area where they can be confined but secure from predators such as foxes, dogs and feral cats. Some houses are equipped with an integral run, which extends under the sleeping accommodation. On a larger scale, it may be necessary to use electrified netting, not only to keep out foxes, but also to control access to new pasture on a rotational basis.
• They need access to fresh, clean ground - either by having alternate runs or by having a moveable house and run.
• A feeder for layer's pellets or mash (powdered form) is needed. Each bird will eat around 130g of this, fed in the morning. Grain (wheat) should be given in the afternoon - scattered on the ground so that they can scratch for it. Alternatively, self-feed grain feeders can be placed at a distance from the house to encourage full use of the pasture. A handful for every three birds is sufficient.
• Provide a drinker so that fresh, clean water is available at all times.
• Provide a dish of grit and crushed oystershell for proper digestion and strong egg shells.
• Clean up spilt food to avoid attracting rats.

- Collect eggs every day and store in cool conditions.
- Eggs may be sold direct to consumers, but only if the eggs are not graded into sizes or described as 'free-range', which is a trade description.
- If eggs are to be sold on a larger scale, a 'packer' (distributor) may be used, or they can be sold direct to retailers. In this case, the producer must comply with the *Egg Marketing Regulations* as well as the stocking densities for *Free Range* legislation. The *RSPCA's Freedom Foods* initiative provides standards for registered, commercial free-range producers to follow if they wish to use this as a description for their eggs. *UKROFS* (referred to in the previous section) has standards for the marketing of organic eggs.
- Don't keep a cockerel, unless you want to annoy neighbours. The old idea that he makes hens lay better is quite untrue. In fact, he should not be allowed to run with layers. The only reason for having one is if you wish to breed your own birds.
- Birds moult once a year, usually in late summer, and often cease laying at this time. Check that they are not affected by lice or mites.
- As daylight hours dwindle in autumn, hens need a little extra light in the house - just enough to extend the day. A 12 volt battery and car bulb will do if the mains supply cannot be extended to the house. Use a timer with a dimmer facility so that the light goes on and off automatically.

Space, Time, Energy and Money

Space - Chickens for a family need little space. Most gardens can accommodate them. On a larger scale, if free-range eggs are to be sold, the space requirements and other stipulations of EC legislation and marketing organisations must be adhered to.

Time - Daily checking, feeding, watering, egg collection, 10-15 minutes (more if enterprise is larger). Weekly cleaning/moving house - 30 minutes.

Energy - Minimal if the flock is a small one. (If you buy a moveable house, make sure you can manage it without problems).

Money - The main starting costs are the house and run or fencing requirements. The birds are relatively cheap. Running costs are low, although feed represents an on-going expense, so make sure you are getting enough back in terms of eggs to make it worthwhile.

Further information

The Poultry Club of Great Britain, 30 Grosvenor Road, Frampton, Boston, Lincs PE20 1DB. Tel: 01205 724081. Looks after interests of pure breeds.
Freedom Foods, RSPCA, The Manor House, Causeway, Horsham, West Sussex RH12 1HG. Tel: 01403 223284. Free-range marketing standards
Sources of Supply: Most housing, equipment, pure breed and free-range suppliers advertise regularly in *Country Garden & Smallholding* magazine. For local availability of hybrids, check the local *Yellow Pages*.
Further reading: Best coverage of the basics - *Starting with Chickens*. Comprehensive detail - *Free Range Poultry*. Both books are by Katie Thear.

Ducks

Ducks are ace consumers of slugs and many people keep them as pest controllers in the garden (after crops have been harvested). They come in all sizes, from the large Aylesbury to the tiny Call duck, so there is room in most gardens, as long as there is a pond. The best egg layers are good strains of Khaki Campbell, followed by Indian Runners, but there are many other varieties with attractive plumage, that are kept for pleasure.

Ducks can be given more freedom in the garden than chickens. They don't scratch, but their flat feet can make a mess of seedlings in a small patch. It's easy to make a small raised pond if you don't have one, and they do enjoy it. They don't naturally go indoors at night, like hens, so you can feed them in the house in the evening to entice them in.

The demand for duck eggs is likely to be on a small, local basis only. They need to be eaten fresh, and the time scale mitigates against their being stocked by the multiples.

If they are to be raised for the table, it's best to concentrate on a commercial strain of Aylesbury or Pekin. If these are bought in as day-olds, they will need to be provided with warm, protected brooding conditions. Check that there is a local demand for them.

Key factors

- Ducks need a pond of sufficient depth to enable them to swim and splash their feathers. It must be cleaned regularly and aerated to avoid a build-up of potentially lethal botulism.
- Despite liking water, they need a house with a ramp entrance to sleep in. Shallow fronted nest boxes will encourage them to lay, although ducks are notorious for laying outside, even in the pond! Keeping them housed until mid-morning normally ensures that they lay inside, before being let out.
- They need to be fed every day - a poultry ration and grain is fine, with the former being given in a wide shallow feeder. There are also waterfowl rations available, as well as rearing feeds for those being raised as table birds. Despite having a pond, they should still be given a drinker with clean water.

Space, Time, Energy and Money

Space: Depends on number kept, but most gardens are suitable, as long as there is a pond. Don't expect to have goldfish and a wealth of pond plants with ducks, unless you provide protection for them.
Time: Daily feeding - 10-15 minutes. Weekly cleaning of house and clearing of pond (if necessary) - 30-60 minutes.

Energy: Minimal, but regular cleaning of the pond can be burdensome, although having an oxygenating airflow from a pump or other source helps.
Money: The biggest costs are the duck house and a pond, if these are not already available, although both can be made as DIY projects. If buying day-olds, a brooder will be needed, although this is a one-off purchase. Running costs are generally low and ducks are not expensive to buy. Their slug clearance work alone is enough recompense for most people.

Further information
British Waterfowl Association, Gill Cottage, New Gill, Bishopdale, Leyburn, N. Yorks DL8 3TQ. Tel: 01969 663693. Looks after interests of waterfowl.
Sources of supplies: Duck breeders, housing and equipment suppliers advertise regularly in *Country Garden & Smallholding* magazine.
Further reading: Ducks and Geese at Home

Geese

Geese are grass eaters and are really only suitable for those with pasture. They also do well in orchards, where they help to keep the grass down, as well as enjoy the windfalls. They don't need as much water as ducks, although an area of water where they can immerse their heads and splash their feathers will be needed. A tank with an automatic supply is suitable.

Geese are long lived - up to 40 years, and mate for life, so if you have a pair, it's a long term commitment. Ganders can be aggressive, so need to be kept away from small children, and anyone else who does not relish being attacked. Conversely, both sexes are excellent watchdogs and prospective burglars are likely to give the site a wide berth.

Popular breeds are Embden, Toulouse, Roman and Chinese. The latter is the best for egg production. If there is a local demand for geese, it may be worth exploring this aspect, particularly if a local butcher will do the killing. The *Humane Slaughter Association* has a useful publication for those killing poultry at home.

Key factors
• Good grass is vital and needs to be available in rotation to prevent build-up of the parasitic gizzard worm.
• Wheat is needed as a feed supplement to grass, particularly after the grass stops growing in autumn. Special waterfowl rations are also available.
• They need a house with a ramp for access, to protect them from inclement weather and from foxes. Despite their size and apparent invulnerability,

they are not safe from the fox!
* Goslings need to be protected against rats in the early stages. It is not true that the parents can see them off!

Space, Time, Energy and Money
Space: Only possible on an orchard or field scale.
Time: Feeding, cleaning house, paddock maintenance and water tank clearance - about an hour a week.
Energy: Not too much required, as long as you can cope with an agressive gander in spring. A dustbin lid makes a good shield!
Money: A house and water tank (particularly one with an automatic supply) are fairly expensive. It may also be necessary to purchase extra fencing. The birds themselves are relatively inexpensive, as long as you don't go in for top show quality ones. Running costs are moderate, although the pasture will require regular feeding, cutting and maintenance.

Further information
The British Waterfowl Association, Gill Cottage, New Gill, Bishopdale, Leyburn, N.Yorks DL8 3TQ. Tel: 01969 663693.
Sources of supply: Breeders and suppliers of housing and equipment advertise regularly in *Country Garden & Smallholding* magazine.
Further reading: Ducks and Geese at Home.

Turkeys

Turkeys are normally bought in as young poults in summer, fattened and then killed before Christmas. If you have a suitable building to house them, they are straightforward to raise. Most turkeys are mass produced and intensively reared - and it shows!

Those raised non-intensively have a much better flavour, and there is usually a local market for quality birds. It is worth doing if you carry out some local market research and can cope with the slaughtering and plucking in December.

There is a choice of quick-growing, white *hybrid* strains or the slower growing *traditional breeds* such as Bronze or Black. The latter are often the choice of the non-intensive rearer. Some people keep a few traditional turkeys as breeding stock and concentrate on selling the poults, rather than on producing Christmas turkeys. These are often breeders whose interest is primarily in the showing sector.

Key factors

- Turkeys need a light, airy building, with access to a sunny outside run which has not been used by other poultry. This helps to prevent the transmission of diseases.
- Poults are normally sold 'as hatched' (A/H) so you will have a mixture of males and females. The stags grow larger than the hens so you will need to take that into consideration when planning your sales. Young birds will need protected brooding conditions until they are hardy.
- Food and water are best provided in suspended feeders and drinkers so that they are not knocked over or have litter scratched into them. Special turkey rations are available, and they will also appreciate grain. Where grain is given, it is necessary to also supply poultry grit so that it is properly digested. Most turkey feeds contain coccidiostats to prevent disease, but it is possible to obtain additive-free feeds from specialist suppliers of quality feeds.
- Turkeys are easily panicked so they should be approached quietly and slowly. They are not aggressive but the fluttering wings are strong, and they may even cause themselves damage.
- Don't under-estimate the amount of work in killing and plucking before Christmas. It may be necessary to employ part-time help.

Space, Time, Energy and Money

Space: A suitable building for the numbers envisaged, and a clean run for access to sunlight. Normally only appropriate for those with farm outbuildings.
Time: Daily feeding and checking - 10-15 minutes. Weekly cleaning of shed - 30 minutes. If rearing for Christmas, allow most of the preceding 3 weeks for non-stop work.
Energy: Minimal until December. Catching, killing and plucking turkeys is hard work
Money: If an outbuilding is available, together with the appropriate feeders and drinkers, the costs are not great. The cost of birds is reasonable, especially if bought 'as hatched'. Feed represents an on-going cost, but the sale of surplus birds and table birds is likely to recoup this. A major expanse is if part-time employees are needed in the pre-Christmas period.

Further information

Sources of supplies: Turkey breeders specialising in traditional coloured breeds, and hybrid suppliers who supply in small numbers will advertise at the appropriate time in *Country Garden & Smallholding* magazine.
Further reading - Profit from Turkeys.

Quail

Quail are classified as game and, being small, can be kept in a small area, including a garden. There are several breeds available in this country, from the small Chinese Painted quail which is often stocked by pet shops for sale to those with aviaries, to the large Bobwhite quail of North America. The Coturnix laying quail is the best for egg and table bird production, with commercial strains being available from specialist suppliers.

Commercially they are kept in cages, but many people prefer to house them in an aviary-type enclosure on humanitarian grounds. They will need extra light to induce them to lay when the daylight dwindles in autumn.

Quail's eggs are luxury items and producers often find a ready market, but as with any enterprise it needs careful research before starting. Special quail egg boxes are available. The table quail market is more for the large scale producer because its labour-intensive nature needs a high turnover to make it economic.

Key factors

• Housing must provide shelter from the elements, and from rats which are their greatest enemy. In a protected aviary make sure there is an area where the little birds can hide from observers. They are shy ground birds and such a place will cater for their instinctive need to take cover.
• They need a high protein ration (being naturally insectivorous) and there are quail feeds available from specialist suppliers. Alternatively, make your own by mixing 1 part chick crumbs: 1 part canary seed: 1 part millet seed. It is possible to buy chick crumbs without coccidiostats from some feed suppliers. Feeders and drinkers designed for poultry chicks are suitable.
• It is possible to rear your own quite easily if you have an incubator. The chicks are only the size of bumble bees, so provide them with shallow drinkers with pebbles, in case they fall in and drown.

Space, Time, Energy and Money

Space: Minimal, unless you are going into it on a big scale.
Time: 5 minutes daily feeding and watering. 15 minutes once a week clearing out the run. Hours spent watching them because they are so interesting.
Energy: Minimal
Money: Low cost enterprise, particularly if carried out on a small scale.

Further information

Sources of supplies: Birds and equipment suppliers are listed at the back of the book *Keeping Quail: A Guide to Domestic and Commercial Management*. This is also the best source of information for anyone who wishes to learn more about most breeds of quail.

Rabbits

Rabbits have been bred for a number of purposes. There are fast growing meat breeds like the New Zealand White and Californian. Angora rabbits are kept for their fine fibre, while a range of breeds are kept as pets and by keen breeders. Whatever the choice, their basic needs are the same. Fresh air, exercise, food, water and protection from predators. If a doe is pregnant, she will need a quiet soft nesting place by herself.

Key factors

• Rabbits need a weatherproof house which can be outside or inside an outbuilding. Commercial meat rabbits are often kept in cages, so that the droppings fall through the bottom. The alternative is a hutch with exercise and sleeping quarters. This will need to be thoroughly cleaned out regularly. During the summer, rabbits will appreciate an outdoor house and run that can be moved around on grass.
• Water can be provided with a nipple drinker attached to the cage. Rabbits can be fed proprietary feeds, but if you are not fattening for meat, provide a range of fresh greens from the garden. Hay is also needed.
• Normally, males and females are kept separate. If you wish to breed your own rabbits, you can introduce the doe into the buck's quarters when she shows signs of being on heat.
• Rabbits are healthy animals and few problems arise if the housing, feeders and drinkers are kept clean. Keep an eye open for ear mites and keep claws trimmed regularly.

Space, Time, Energy and Money

Space: For rabbit farming or serious breeding, you will need a large building, but if you are keeping just a few, then little space is needed.
Time: Rabbits need twice daily replenishment of food and water, with a daily check over and a twice weekly clean out.
Energy: Rabbits need little exertion; large buck rabbits need firm handling.
Money: Minimal, unless planning a large enterprise or specialised stock.

Further information

British Commercial Rabbit Association, 41 Broad Street, Syston, Leicester LE17 1GH - offers advice for anyone setting up a commercial rabbit enterprise.
British Rabbit Council, 7 Kirkgate, Newark, Notts NG24 1AD. Tel: 01636 76042 - advice on keeping rabbits, plus breeding and showing exhibition rabbits.
Angora Fibre Centre, Gillian Sharp, 16 Braithwaite Road, Keighley, West Yorks BD22 6PA. Tel: 01535 602033 - advice on keeping Angora rabbits for fibre and on collecting fibre for processing
Further reading: Rofe's *Pre-Start Advice for Rabbit Enterprises* - meat or fibre production. *Rabbits* - complete pet owner's manual. *Dwarf Rabbits* - complete pet owner's manual

Sheep

To keep sheep you need land and good pasture. There are more than 50 breeds to choose from, as they have been developed for different climatic conditions and terrain. Hill breeds are generally smaller and more hardy than lowland sheep. They also have very different wool characteristics for use in everything from the finest wool garments to heavy duty carpets.

Sheep produce both meat and wool, so you need to be clear why you want to keep them. It may be that you simply wish to 'use up' an area of grass and produce a couple of lambs for the freezer. Many small flock owners keep coloured breeds for the fleece which can be spun into natural coloured wool for making up into garments. These, in turn, can represent a source of value-added income.

Sheep need to be protected against a variety of diseases requiring vaccination. Because of this, and the potential problems associated with lambing, it is highly recommended that anyone considering keeping sheep attends a sheep management course.

Sheep fit in well with free-range poultry. They eat the grass down so that small new growth is made available for the birds which cannot cope with taller growth.

If there are enough ewes to warrant it, it may be a good idea to have a ram, but be aware of potential aggression. A ram can, and does, butt!

Whatever the scale of operations, every sheep owner is now required to register the flock with the *Ministry of Agriculture*. A *Movement of Livestock Record Book* should also be kept, indicating when animals were moved on or off the premises.

Key factors

• Good fencing is essential. Some of the more primitive breeds are very agile at climbing. Electrified fencing can be used to reinforce existing boundaries as well as to control rotational grazing.
• Check that pasture is in good heart, with no poisonous plants such as ragwort. It will need to be made avilable in rotation, with periods of rest and feeding. Exhausted pasture may require reseeding.
• Unless sufficient hay can be made on site, it will be necessary to buy in supplies. Hay is required when the grass stops growing. Sheep also need concentrates with appropriate mixtures for pregnant ewes and lambs.
• A barn or other covered area is needed to store the hay bales.
• An outbuilding or adapted polytunnel is required for shelter, particularly

at lambing time. This needs to be equipped with pens. These can be made from portable hurdles which are also a boon in general management and control of sheep.
• A supply of fresh, clean water is essential. The most efficient way of supplying this is to have a field tank which automatically fills by means of a ball-cock attachment. This will need regular checking and clearing.
• There is a cycle of tasks associated with keeping sheep: caring for the pregnant ewes, lambing, vaccinating, worming, rearing lambs, summer health watch, shearing, tupping, selling meat lambs or having them slaughtered. Many flock owners also show their prize animals at agricultural shows. In fact, visiting a show where many sheep are exhibited is an excellent way of viewing the different breeds and talking to breeders.

Space, Time, Energy and Money

Space: Several acres required, as well as suitable shelter and good pasture.

Time: Fairly time-consuming. Daily attention to feeding and health. Periodic attention to health matters, shearing, foot trimming, and lambing. Daily watch in the summer in case of blowfly attack.

Energy: Fairly strenuous. You need to be able to handle sheep for foot-trimming, etc, so if there is a problem, choose a small breed and ensure that you have help.

Money: Capital needed for buying good stock initially, as well as any equipment such as hurdles, watering system, electric fencing, field shelters, etc. Hay-making equipment may also be needed, although this is generally available secondhand at quite reasonable prices at farm sales. Shearing equipment may be needed, unless it is being done by a contractor. Feeds and medical supplies represent an on-going cost. Keep an accurate record of income and outgoings to ensure that you are at least breaking even. Contact the local branch of *MAFF* to see whether you are eligible for any subsidies.

Further information

National Sheep Association, The Sheep Centre, Malvern, Worcs WR13 6PH. Tel: 01684 892661.
Rare Breeds Survival Trust, NAC, Kenilworth, Warks CV8 2LG. Tel: 01203 696551. Looks after the interests of rare breeds of sheep and other stock.
Further reading - Introduction to Keeping Sheep - Ideal first book by Upton and Soden. *Sheep Ailments* - useful picture guide
Sources of Supplies - Breeders' Directory in *Country Garden & Smallholding* magazine for stock. Equipment from mail order advertisers or local farm suppliers.

Goats

Goats are herd animals so at least two need to be kept. There are different types of goat developed for different purposes. Most goats are dairy goats kept for their milk and there are a number of breeds available such as the Saanen, British Saanen, Toggenburg, British Toggenburg, British Alpine, Anglo Nubian, Golden Guernsey and English. There are also Boer goats kept for meat, Angora and Cashmere goats for their fibre and Pygmy and Bagot goats kept mainly for pleasure. (Fibre goats are kept in a similar manner to sheep). If you are considering keeping dairy goats, don't underestimate the time spent in milking and dealing with the milk

Goats are intelligent animals and full of interest. They respond well to having a little extra care and attention. They will, for example, enjoy having some large logs or boulders on which to jump if they are in a confined run.

Courses are available in goat management and dairy practice, and it is recommended that the prospective goatkeeper attends one of these before starting. As with sheep, there are regular management tasks such as vaccination, foot trimming, etc.

All goatkeepers, regardless of the number of animals thay have, are required to register them with the *Ministry of Agriculture*. A *Movement of Livestock* record is also required, recording any movements on or off the holding.

Key factors

• Dairy goats need regular mating to produce kids and a flow of milk. A good milker can produce a gallon every day at her peak, so you will need to make something with the surplus. Yoghurt is the usual choice, together with soft cheese. If any products are to be sold, the premises will need to comply with the appropriate dairying and food handling regulations. (Contact *MAFF* for details).

• Goats are browsers and are often more interested in the content of overhanging tree branches and boundary hedges than the food beneath their feet. Make certain that saplings are protected, and fences and gates are escape proof. (Intelligent goats can often work out how to open a gate!) The effect of goats browsing on large leaved weeds, such as docks, improves the overall pasture over a period of time. Pasture should be made available on a rotational basis, and rested and fed when not in use. Poisonous plants such as ragwort should also be removed.

• Goats can be kept in an exercise yard, but this entails the owner bringing

a regular supply of green food to keep them fed and interested. As with all milking livestock, they can drink a great deal and fresh water needs to be checked and replenished regularly.
• They need a regular diet of concentrate feed (special goat rations are available), hay and browsing, as referred to earlier.
• Goats normally need milking twice a day. This can be by hand or with the aid of a mobile machine milker.
• Mating can be carried out by AI or by a trip to visit a billy at a nearby stud farm. Check that it is a CAE accredited unit, and that any goats purchased are also from such a herd, free of CAE disease.

Space, Time, Energy and Money

Space: To keep dairy goats you will need an outbuilding that can be adapted into goat pens, a milking parlour and a feed/ hay store. You will also need the space at home for dairying activities. Outside, you will need well fenced land or an exercise yard. Some use the latter during the winter months to rest the pasture.
Time: Milking twice daily and dealing with the milk is obviously time-consuming, with the amount of time depending on the scale of activities. Replenishing water and other daily checks can take another 30 minutes. Other jobs are weekly or less.
Energy: Keeping goats is a physical activity, but it needn't be strenuous if you are well organised. Female goats are easy to handle and respond positively to good care. Only consider having a male if the scale of breeding operations warrants it. He will need his own quarters and yard.
Money: Starting up costs will vary depending on what needs to be done. Purchase of stock, milking and dairying equipment new is quite high. Running costs include buying hay, concentrates, bedding and health products.

Further information

The British Goat Society, 34-36 Fore Street, Bovey Tracey, Newton Abbot, Devon TQ13 9AD. Tel: 01626 833168.
British Angora Goat Society, 4th Street, NAC, Stoneleigh Park, Kenilworth, Warks CV8 2LG. Tel: 01203 696722.
The British Boer Goat Society, Eaglemead Farm, Longford Road, Thornford, Sherborne, Dorset DT9 6QN. Tel: 01935 872234.
The Pygmy Goat Club, 7 Chapel Road, Rhiwceiliog, Pencoed, Mid-Glam CF36 6NN. Tel: 01656 860091.
Supplies - Local goat club can provide advice, sources of local stock, equipment and feeds. There are also good mail order suppliers throughout the pages of *Country Garden & Smallholding*.
Further reading - Home Goat Keeping. A brief overview for intending goatkeepers *All About Goats*, and *Goatkeeper's Veterinary Book* are both useful manuals for goatkeepers.

House Cow

A cow needs a considerable area of good pasture land. There should also be a field shelter, and fences and gates should be in good order. On a small scale the most suitable breed is the Jersey, although some smallholders keep a Guernsey or the smaller Dexter. There are also many other breeds available; an agricultural show is a good place to see the range.

When buying a cow or calf, try to ensure that it comes from a herd that has been grass-fed, and has not been given concentrate feeds containing animal protein in the past. This is very difficult to establish, of course, but you need to be aware that it was the practice of using diseased animal protein in cattle feeds which created BSE in British herds. There is no evidence of BSE being transmitted through milk or dairy products.

The gentle Jersey produces creamy milk at only half the rate of a Friesian, but that can still be as much as 2 gallons a day for the household. Some of the surplus skimmed milk (after the cream has been removed) can be used to raise pigs. There is also a range of dairy products that can be made at home, including butter, cream, cheese, ice cream and yoghurt.

Key factors
- A cow needs a well ventilated building or field shelter to protect her from the weather, and during calving.
- There should be a separate milking parlour and feed store.
- Make sure that the pasture is in good heart with no poisonous plants, and that fences and gates are in good repair.
- Pasture will need to be made available in rotation, rested and fed. If you make your own hay, the necessary equipment will be needed.
- During the winter months there is little nutrition in the grass and pasture can be left to rest whilst the cow has hay, fodder crops like kale and beet, or silage (a kind of pickled grass) if available.
- To keep your cow in milk she will need to be mated regularly. This can be done through AI. If she is crossed with a suitable beef breed, the calf can be grown on for the freezer, or for sale, but bear in mind the comments made earlier about BSE.
- A house cow will need milking twice a day, and home dairying - making use of the surplus milk, will need to follow on.
- As with all livestock, you will need to watch her health. See that she isn't bothered by flies, and that her feet are in good condition Like all pasture animals, she will need regular worming.

- When milking, follow all the hygienic requirements and be sure to check that the milk is free from taints or clots which might indicate mastitis. It needs veterinary attention and treatment straight away.
- A house cow must be registered with *MAFF*, and a *Movement of Livestock* record book kept to indicate all movements on and off the holding.

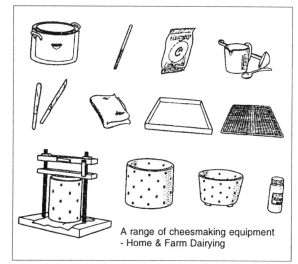

A range of cheesemaking equipment - Home & Farm Dairying

Space, Time, Energy and Money

Space: The cow needs plenty of space, at least 2 acres of good pasture, and you are likely to need to buy in at least part of your hay. She needs suitable indoor quarters, a milking parlour and a feedstore. A robust outbuilding can be adapted for her use. Milk storage and dairying activities will also need their own separate space. For home use this will probably be the kitchen and larder. If you plan to sell any of your dairy produce, the dairy will need to meet the appropriate dairying and food handling regulations. (Details from the local office of *MAFF*).

Time: Milking and dairying will probably take up to 2 hours a day at the peak of lactation. Other daily tasks - around 10-15 minutes. More time is needed at specific times such as calving.

Energy: Carrying hay, straw and feed sacks can be tiring, as can mucking out. Most other tasks need not be strenuous.

Money: Cost of cow, preparing fencing and outbuildings can be high. Purchasing milking and dairying equipment adds to this. Ongoing costs include hay, straw, concentrates and animal health products.

Further information

Relevant cattle breed societies can supply sources of breeders. Local farm contacts also bring knowledge of local sources of equipment and supplies. Mail order suppliers are in *Country Garden & Smallholding* magazine.
Further reading - *Caring for Cows* - a book for house cow owners
Home and Farm Dairying - milking, dairying practice, making butter, cheese, ice cream and yoghurt at home. Suitable for all types of milk - cow, goat and dairy ewe.

Pigs

Pigs are normally kept for meat, and will fatten rapidly, particularly if there is surplus whey available from dairy animals. You can buy in young weaners in spring at between 3-8 weeks old, and they can be reared for 4-6 months until they reach slaughter weight. There is a break over the winter, then the cycle begins again when the milk surplus is growing.

Instead of buying weaners you can have a breeding sow and rear your own piglets, mating her by artificial insemination. There are a number of breeds popular with smallholders, including the Gloucester Old Spot, Berkshire, Tamworth and Oxford Sandy and Black. There are also the miniature breeds like the Vietnamese Pot Belly and the Kune Kune. These can be easier to handle on a small scale, and kill out at about half the weight of the full size breeds. If you decide to breed from the small breeds, remember that they continue to grow after slaughter weight and finally mature at quite a large size.

Don't have a boar unless you intend to go in for serious breeding. He will need his own quarters and yard. Like most male livestock, he can be dangerous and needs competent and skilled handling.

It is a good idea to attend a practical course on pigs before starting to keep them, so that all aspects of health care and management are familiar.

All pig owners are required to register their animals with the *Ministry of Agriculture*, and to maintain an up-to-date *Movement of Livestock* Register.

Key factors

- Pigs are ideal ploughs, rooting up weeds and turning over the ground, while they manure it at the same time. They need to have a house and temporary fence (eg, electric netting) to confine them to the area. As soon as it has been cleared, move the house and fence ready for the next section.
- A strong field shelter is needed, or if you haven't this amount of space available, a house and run, ideally block built with a concrete floor will be suitable. Good drainage is essential for this so that it can be regularly hosed.
- Pigs are strong and can find weak points in fencing. Use pig netting, well secured so that they cannot get under it.
- Weaners can be fed on weaner meal mixed with surplus whey if you have it. They will also consume garden surplus fruits and vegetables. There are also concentrate feeds for adult pigs.
- Good health care is required, including regular worming and foot care.
- Pigs can be slaughtered at home, provided that it is done according to

the correct standards and that none of the meat is to be sold. It should be done only by those who are experienced in the process. Follow the advice of the *Humane Slaughter Association*, referred to earlier. Alternatively, a registered slaughterhouse will slaughter and joint the pig, but there is the stress of having to transport the animal to the site.

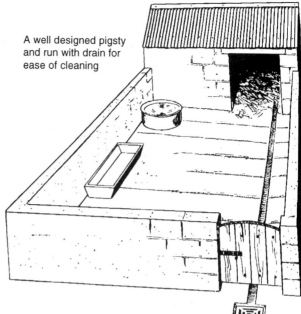

A well designed pigsty and run with drain for ease of cleaning

- You can have a go at curing your own hams, or have them done commercially. Home-made sausages are also fairly easy to make, with casings and equipment available from specialist suppliers.

Space, Time, Energy and Money

Space: Depends on the number you keep, but a house and run for a couple of pigs can be around 30 square metres.

Time: Feeding takes time if you are mixing up each time - up to 40 minutes a day. Cleaning of a sty run amounts to around 15 minutes a day. Pigs are clean animals and will not foul their sleeping area.

Energy: Carrying feed and cleaning out is physical work.

Money: High initial cost if you have to prepare a house and strong fencing, and purchase stock. Day to day costs arise from feeds and maintaining good health.

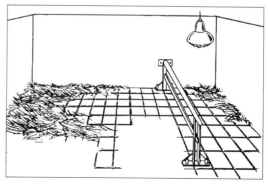

Pig farrowing area in an outbuilding, with a warm place for the piglets behind the crush bar

Further information

Sources of supplies - Breeders' Directory in *Country Garden & Smallholding* magazine for pigs

Further reading - Practical Pig Keeping

Bees

Successful beekeeping needs knowledge and experience, not to mention an investment in equipment and tools. There are local authority courses available in many areas, as well as local branches of beekeeping associations. Joining such a society is an excellent starting point because it enables you to 'apprentice' yourself as helper to a local beekeeper. From this, you will be able to establish whether beekeeping is for you. It is not feasible for everyone, particularly if a member of the family is allergic to bee stings!

The benefits of beekeeping are considerable - good pollination levels for fruit trees and crops, and home-produced honey. Be aware, however, that Varroa disease is now in the UK, and precautionary measures need to be taken against the mites that carry it.

Key factors
- Beehives take up little space, but they need careful siting away from paths and people, not under trees and protected from strong winds.
- You will need at least two colonies to begin with and they can be sited at home or elsewhere, eg, moved to commercial orchards, or the moors for heather honey.
- Before buying bees or equipment, take local advice as second hand items may be available. These should be fumigated before use in order to minimise the risk of disease transference.
- Much of the beekeeper's work takes place between May and September, culminating in the honey harvest.

Space, Time, Energy and Money
Space: Little space is needed, just an appropriate site.
Time: Most tasks take about 10-15 minutes during the summer months, as long as you take precautions to avoid swarming. Extracting the honey crop takes about 2-3 hours a hive.
Energy: Minimal unless you are involved with transporting hives of bees to other locations, or need to recapture a swarm.
Money: High set up costs if everything has to be purchased new. Needed are hives, bee colonies, protective clothing, beeswax, hive tool and smoker, extractor, etc. Running costs are minimal, unless honey jars, labels etc, need to be bought every year. Surplus honey (if you can bear to part with it) can be sold locally without too much difficulty.

Further information
Supplies - information on bees and equipment through local beekeepers and bee magazines.
Further reading - *Beekeeping for Beginners* and *Background to Beekeeping*. Both good beginners' books from *Country Garden & Smallholding* booklist.